INTRODUCTION TO EARTHQUAKE RETROFITTING

INTRODUCTION TO

EARTHQUAKE

RETROFITTING

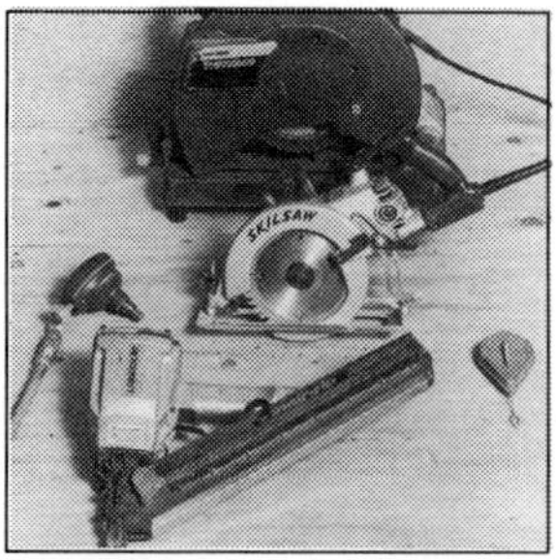

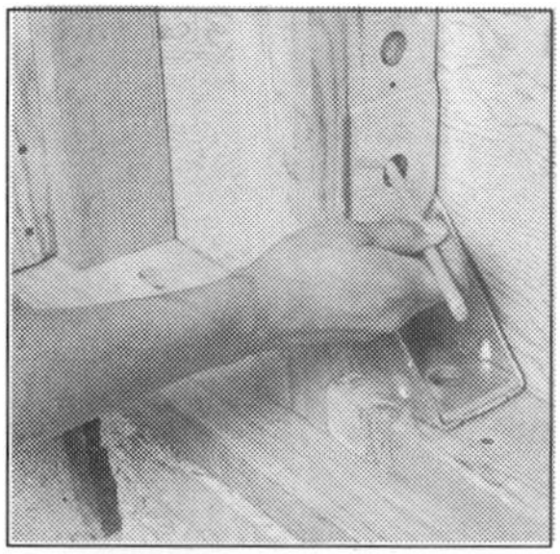

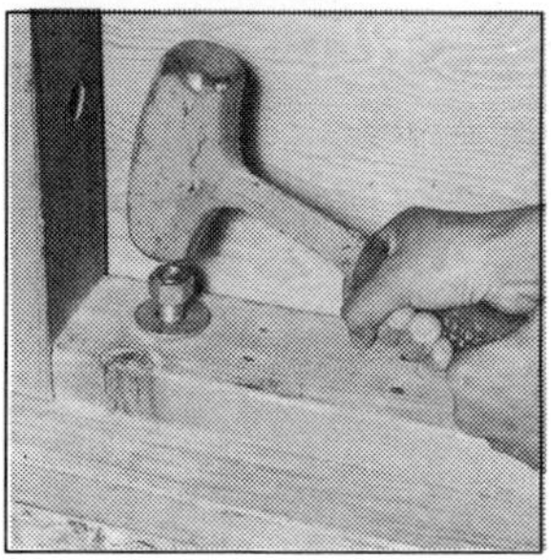

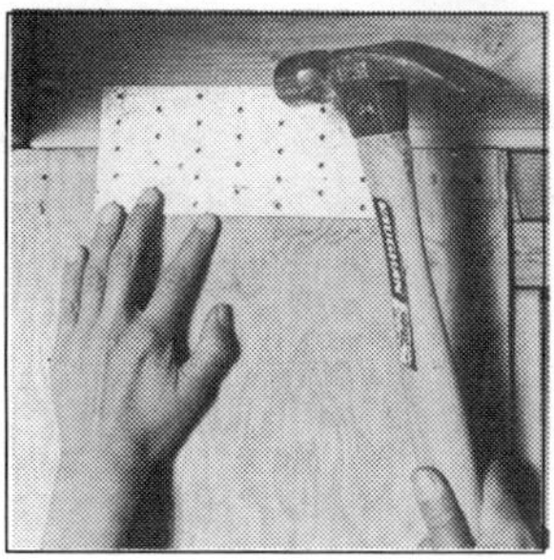

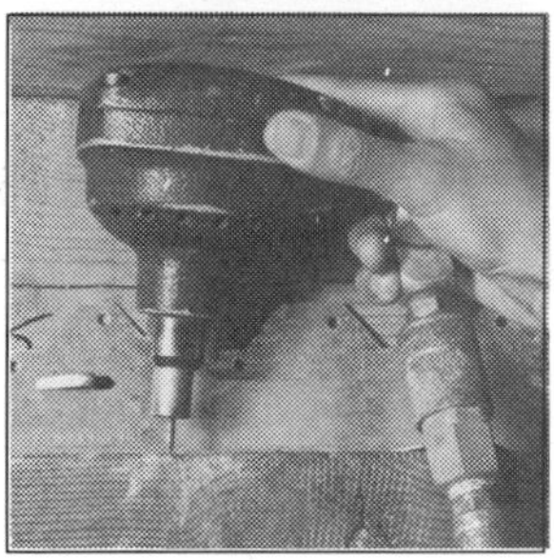

TOOLS & TECHNIQUES

ACKNOWLEDGEMENTS

Introduction to Earthquake Retrofitting 2nd Edition

Published by The Building Education Center, Berkeley, California

Written by Charles Smith and Cindy Furukawa

Photography by Pacific Technical Services, San Francisco, California

Design by Zibra Design Group, San Francisco, California

Printing by Regent Press, Oakland, California

This book was printed on recycled paper.

Special Thanks to:

James Gillett

Tony DeMascole

The tools and hardware shown are representative of professional quality tools used by experienced contractors and do not represent an endorsement of any particular manufacturer or model.

Manufactured in the United States of America

ISBN # 0-963281-80-1

INTRODUCTION

The purpose of this manual is to introduce common tools and techniques used in residential seismic retrofitting.

The key principle in seismic strengthening work is to think of a house as a series of interlocking parts that must be tied together into one complete unit. A common error is to assume that bolting the mudsill (the wood piece that rests on the concrete) to the foundation is sufficient. This is incorrect. If the walls of the house are not adequately tied to the bolted mudsill, the lateral or side to side movement may cause the house to collapse. The most vulnerable parts of the house are the walls between the foundation and the first floor - called cripple or knee walls. It is not enough to strengthen just one part of the house - all parts must be tied together.

This manual illustrates the most common techniques used in seismic retrofitting:

- ➢ Bolting the foundation to the mudsill (expansion bolts and threaded rod/epoxy)
- ➢ Tying the mudsill to the cripple walls (plywood shear walls)
- ➢ Tying the cripple walls to the floor framing (blocking and hardware ties)

Because every home is unique, you need a structural engineer who is familiar with wood frame construction to specify exactly what connections and hardware you need before you start any retrofitting work. The information provided in this manual is designed to provide general consumer education to homeowners, not to substitute for professional engineering or guide work performed on a specific home.

The framing and foundation must be structurally sound. Repair all dryrot and termite damage before beginning any retrofitting.

TABLE OF CONTENTS

TOOLS AND EQUIPMENT

Seismic strengthening work goes much faster and easier with the right power and hand tools.

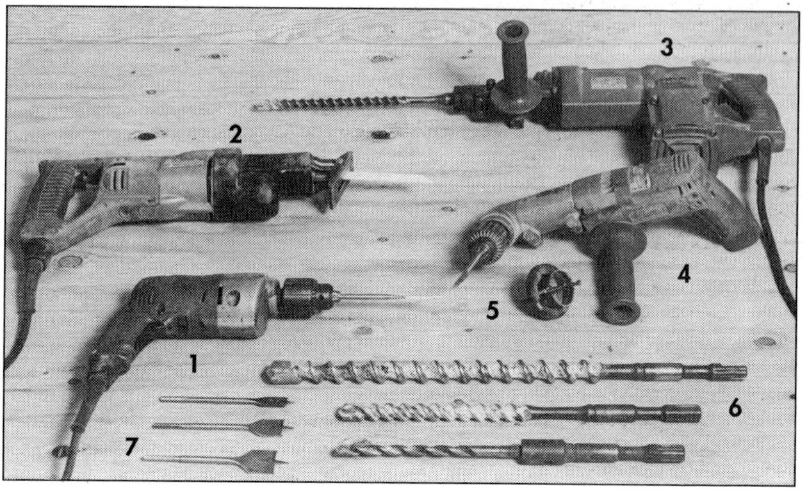

➢ POWER TOOLS

The tools shown are professional quality. Good tools make any job easier.

1. 3/8" ELECTRIC DRILL

Drills designed with a detachable handle are safer because you can control the tool better when you have a secure grip. A drill can be violently twisted out of your hands if the bit binds or jams. See page 12 for an example.

2. RECIPROCATING SAW

Milwaukee Tools' "Sawzall®" has become a generic name for this type of saw.

3. ROTARY HAMMER

This tool has a simultaneous drilling and hammering action that is designed for drilling through concrete. While a heavy-duty drill without the hammering action can do the work, it will take much longer. This can be rented, along with the appropriate bits, from tool rental firms.

4. 90-DEGREE DRILL

Retrofitting work often requires drilling holes in tight places. This drill is required when there isn't enough clearance for positioning the drill.

BITS

5. HOLE SAW BIT

These bits are used to drill 1 1/2 to 3 inch diameter ventilation holes in plywood shear walls.

6. MASONRY CARBIDE BITS FOR THE ROTARY HAMMER

You will select the bit to fit the diameter and length of the bolt your engineer specifies. These bits are expensive ($60 and up each) so they should be handled with care.

7. WOOD BITS

Flat wood bits are used with a standard drill to bore out the wood mudsill and studs for bolts.

1. METAL CUT-OFF SAW

These saws use abrasive blades to cut through steel rods.

2. CIRCULAR SAW

This particular portable saw is a heavy-duty professional quality tool; you can also use a household circular saw.

3. NAIL GUN

This compressed-air gun drives an 8d to 16d nail in one powerful stroke. The nails come in strips and are more expensive than regular nails.

4. PALM NAIL GUN

Only one nail can be inserted at a time in this tool. It drives each nail in a series of short consecutive strokes. Note the air hose which connects the tool to the compressor.

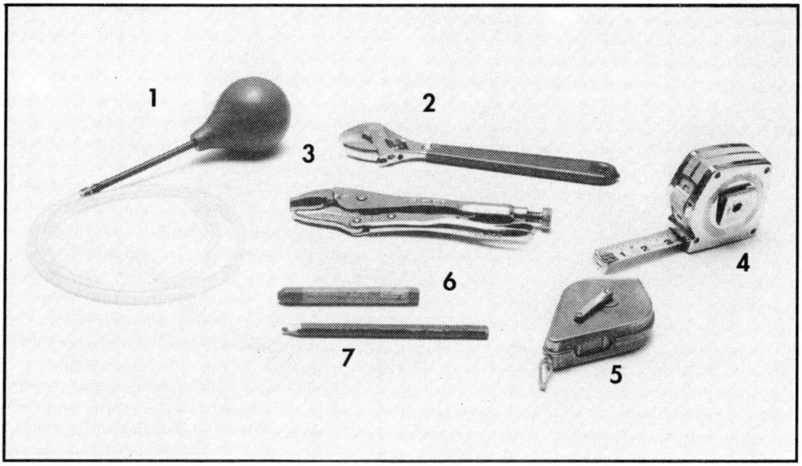

➤ HAND TOOLS

1. PLASTIC TUBING

The flexible rubber ball attached to the tubing blows concrete dust out of holes drilled for bolts.

2. ADJUSTABLE WRENCH

3. VISE GRIP PLIERS

4. TAPE MEASURE, 20' OR 25'

5. CHALK LINE

6. CARPENTER'S CRAYON (KEEL)

7. CARPENTER'S PENCIL

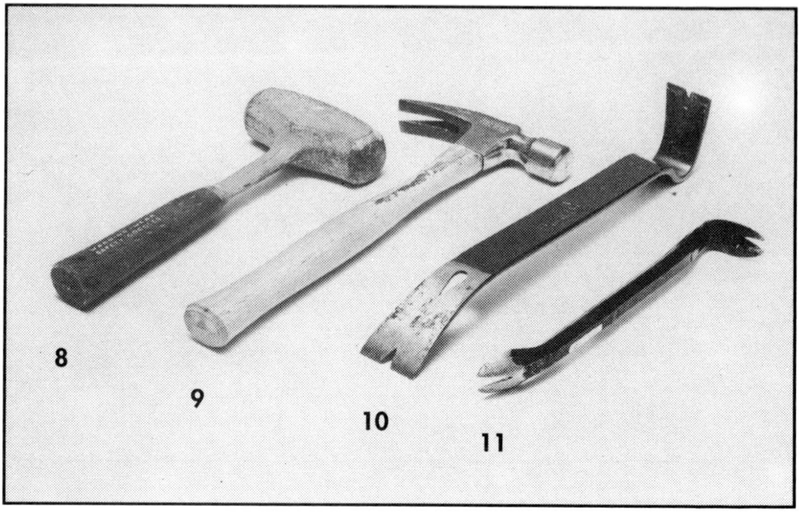

8. FOUR POUND SLEDGE HAMMER

Regular hammers are too light for installing expansion bolts.

9. HAMMER

Hammers are sized by the weight of the hammer head. This is a 20-ounce framing hammer. Twenty to 24-ounce hammers are more useful than 16-ounce finish work hammers which are too light for retrofitting work.

10. PRY BAR

11. "CAT'S PAW" NAIL PULLER

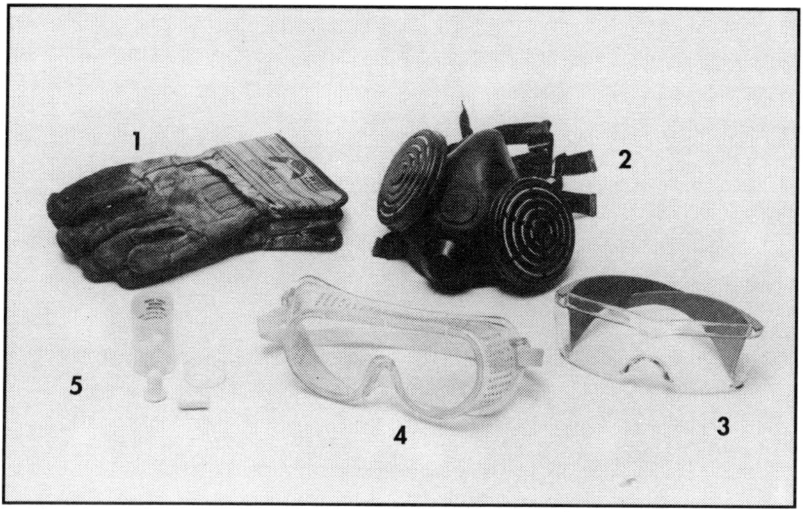

➢ SAFETY EQUIPMENT

Always wear proper safety equipment when doing retrofitting work.

1. GLOVES
2. SAFETY GLASSES
3. SAFETY GOGGLES
4. RESPIRATOR
5. EAR PLUGS

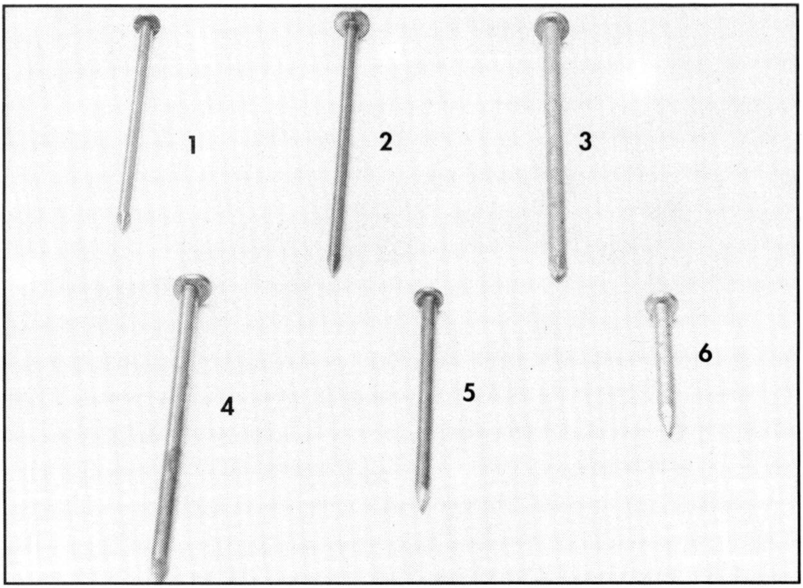

➢ NAILS

A variety of nails are used in retrofitting work. The letter "d" used to denote nail length is from the Roman word for penny, denarius; in Old England, nails were priced by how many pennies it took to buy 100 nails. 16d is pronounced "sixteen penny;" a modern 16d nail is 3 1/2" long.

1. 8d COMMON NAIL

2. 16d BOX NAIL

Note the thin shank. Do not use these nails in retrofitting.

3. GALVANIZED 16d COMMON NAIL

Galvanizing protects nails from rusting; galvanized nails are used in areas exposed to dampness such as basements and roofs.

4. 16d COMMON NAIL

5. "SINKER" 8d NAIL

These nails are coated to make them easier to drive.

6. JOIST HANGER NAIL

Teco is one of several manufacturers which makes these thick-shanked, short nails which are designed for use with metal clips and ties.

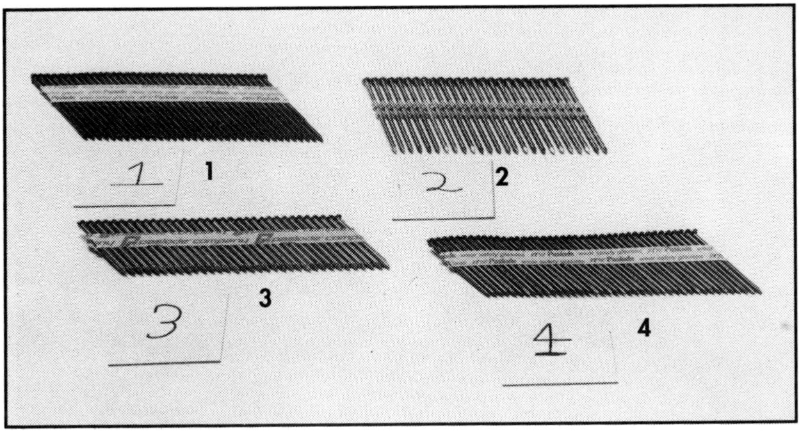

GUN NAILS

These strips of nails are held together by glue or tape, and are manufactured to fit particular brands of guns. They come in a variety of sizes and types for different applications. Here are four common types.

1. 16d COMMON NAILS
   This size is typically used in rough framing carpentry.
2. 16d COMMON NAILS
   This is a different manufacturer's nail strip.
3. 8d COMMON NAILS
4. 10d RING SHANK
   These nails are designed for use on plywood shear panels.

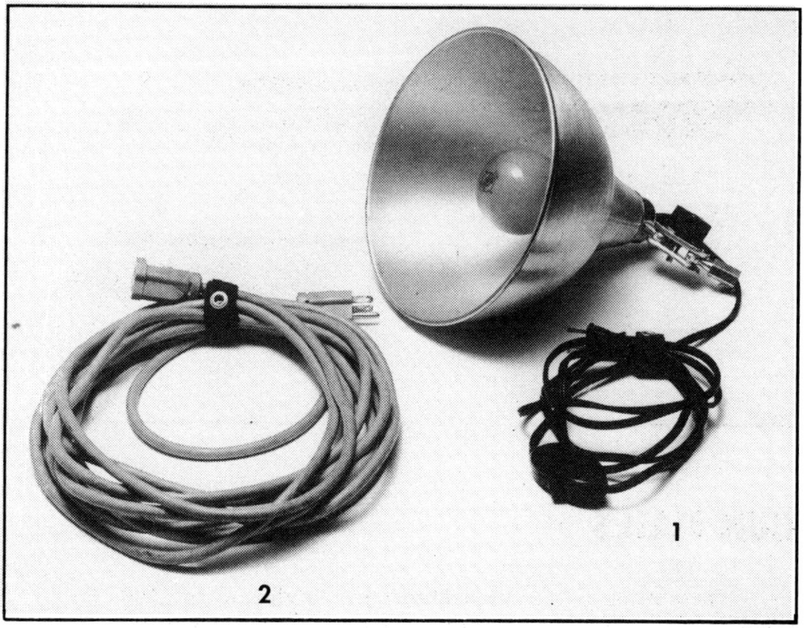

➢ LIGHTING

Good lighting is an important safety feature for retrofitting work.

1. SHOP LIGHT
2. 12 GAUGE EXTENSION CORD

BOLT INSTALLATION

Earthquakes cause damage in two ways: lateral movement and uplift. Bolts resist the lateral movement, preventing the house from shaking or sliding off the foundation. ❧

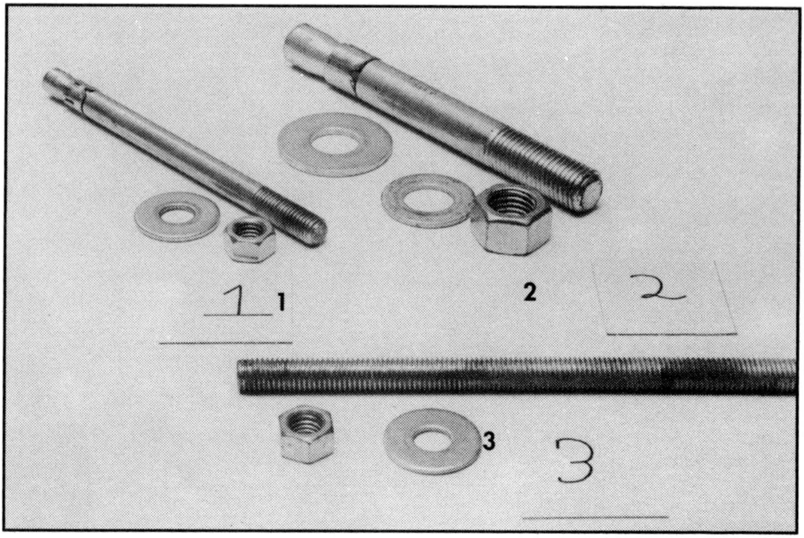

1. 5/8" EXPANSION BOLT.

The collar on the bottom of the bolt expands as you tighten the nut, wedging the bolt tightly in the hole. The hole must be the same size as the bolt, and the foundation must be good quality, hard concrete for these bolts to work effectively. If the concrete is too soft, they may pull out when you tighten the nut. These bolts install quickly.

2. 1" EXPANSION BOLT.

Both bolt length and diameter increase with the size and height of the building. Bolts come in 1/8" increments: 5/8", 3/4" and 1"; for residential structures, 5/8" and 3/4" are most common sizes.

3. THREADED ROD.

A more time-consuming and expensive option is using threaded rod and epoxy adhesive. Typically, adhesives are used when the concrete is too soft or weak for expansion bolts. If you're unsure, have an engineer inspect the foundation.

If your mudsill is wider than the studs, you need to add 2 X 4 blocks in between the studs to provide a perimeter nailing surface for the plywood shear wall that you'll be installing later.

Blocks should fit tightly. Use 12d or 16d common nails, about 3" apart, with a minimum of 6 per block. To avoid splitting the wood, you can pre-drill some 1/8" pilot holes. Start the nails at least 1-1/2" in from the ends.

Many professionals use termite-resistant pressure treated lumber in the foundation area.

Follow your engineer's specifications in marking bolt location. In this case, the bolts are spaced 4' (48") apart. You need to install bolts at each end of every wall, and on either side of any joint or step in the mudsill. These bolts should be between 4" and 12" from the end of the mudsill. It's easier to see your marks if you use a light-colored carpenter's crayon (keel).

Your engineer should specify both the bolt's length and diameter; the bolt should go at least 5" into the concrete. With a 2 X 6 mudsill and a 2 X 4 block (each 1-1/2" thick), you would need at least an 8-1/2" long bolt to go through both pieces of wood and penetrate 5" into the foundation. Remember to leave at least 1/2" of threading for the washer and nut.

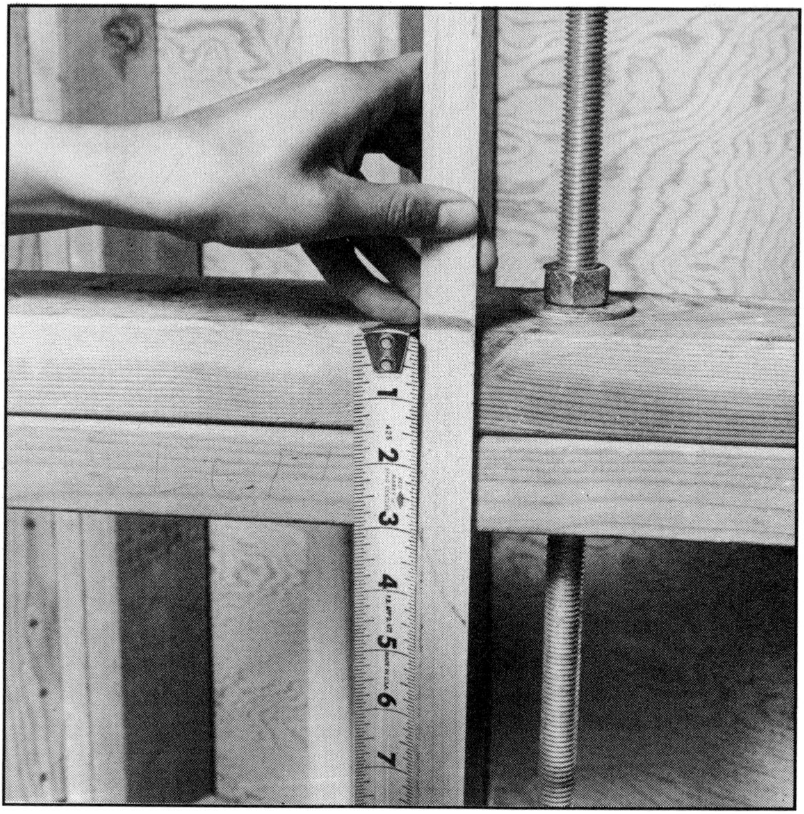

A marked stick makes it easy to check if your hole into the foundation is deep enough for the bolt. If you're using expansion bolts, you can drill each hole a bit deeper. Note that the stick is marked at the top of the block.

Drill through the blocking and mudsill with the wood bit. To avoid drilling too deep and hitting the concrete, you can mark the bit with a piece of masking or electrical tape. Note that the drill is held securely with both hands.

Use a respirator, ear and eye protection when using the rotary hammer drill; tie back your hair, too. Keep the drill vertical, and don't lean heavily on it, just hold it firmly and allow the tool to do the work. Some hammer drills have depth gauges; adjust the depth gauge as you would the stick, or use the marked stick. When using expansion bolts, use the same diameter bit as that of the bolt. When using epoxy, make sure the hole is at least 1/8" larger than the bolt diameter. Drilling into concrete creates a lot of dust.

The drill bit may become wedged if it hits one of the steel reinforcing rods or aggregate rocks embedded in the concrete. If this occurs, try to free the bit by pushing down slightly and then pulling it out slowly. If this doesn't work, turn off the tool and pull up on the handle. As a last recourse, release the bit from the tool and use either a pipe wrench or a vise grip to turn the bit counter-clockwise. Set the wrench below the top of the bit that fits into the drill so you won't damage the bit. This action may back the bit out of the hole. If none of this works, you have a $75 anchor bolt in your foundation.

Check the depth of the hole. This hole needs to be deeper. Note the nailing pattern on the block.

Most of the dust will be removed as you withdraw the bit from the hole, but you need to clean the hole out thoroughly. Here, we are using 3/8" plastic tubing and a rubber bulb. For photographic clarity, we used a 36" length, but to avoid getting dusty, use a longer piece.

To protect the threads of the bolt from being damaged, put the nut on so that a few threads inside the nut are exposed, i.e. the nut is above the bolt. Make sure the washer is in place before inserting the bolt. You'll have to hit the bolt pretty hard to drive it into the hole. While a regular hammer will do the job, it's faster to use a four pound short handle sledge hammer as depicted.

Tighten the nut with either a crescent or socket wrench. The bolt is properly tightened when the washer crushes the wood slightly. Once tightened, don't overdo it. Over-tightened bolts will spin, much like when you overtighten a bolt or screw and strip the threads.

This is a properly installed bolt. Note that while it could have been installed between 4" and 12" from the corner, for ease of installation it was put in the larger bay.

EPOXY

Epoxy adhesives and threaded rods are an alternative to expansion bolts. They are more expensive and time consuming to install, but they are more appropriate for poor quality concrete. Epoxy can be purchased in two-part bulk mixtures or individual capsules. We illustrate the less expensive two-part type. Some epoxies are stronger than others; follow your engineer's specifications. ❧

Cut threaded rod to desired length with a metal cutting blade in a circular saw, as shown above, or in a sawzall, as shown below. When using a sawzall, secure the threaded rod in a vice (do not use the chop saw as a vice).

Use eye protection when using metal cutting blades; abrasive blades generate showers of sparks.

Follow manufacturer's instructions carefully when handling, mixing and disposing of epoxy adhesives. These are toxic products. Wear gloves and prepare the mixture in a well ventilated area. Mix only as much as you can use in a short period of time.

To insure a good bond when using epoxy, make sure the hole is clean. Dust will interfere with the adhesion. Fill the hole completely.

Embed the threaded rod so that some of the epoxy oozes out from the top. Note that there is at least 1" of threaded rod exposed for the washer and nut. It's okay if it's longer.

Allow the epoxy to cure for at least 48 hours before installing the washer and nut.

HOLDOWNS

While bolts resist lateral movement, holdowns keep the building from lifting or tilting off the foundation. Your engineer's drawings will show you where to place these. Typically, they are installed on both ends of each structural shear wall. ❧

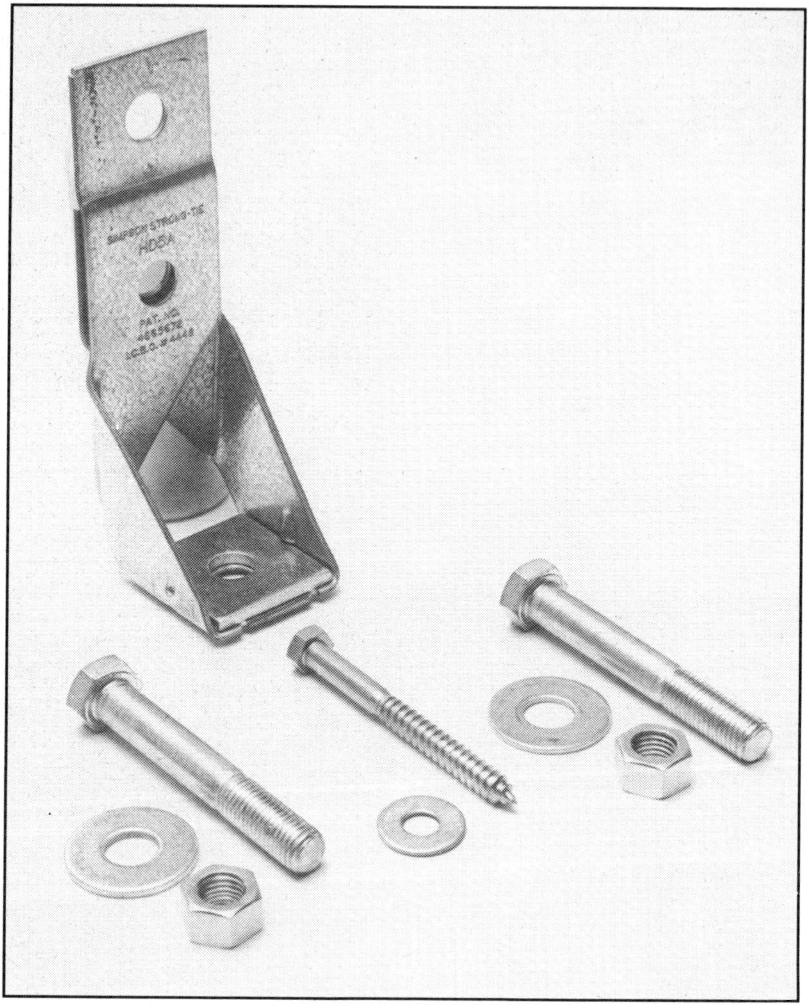

➢ HOLDOWN AND BOLTS

The hold down is a heavy metal bracket connected to the foundation with either an expansion bolt or threaded rod, and to the framing with machine or lag bolts. Machine bolts go through a double stud or post and are fastened with a washer and nut. Threaded lag bolts are used where it is not possible to install machine bolts.

Holdowns must be connected to either a double stud or a 4" x 4" post; a single stud is not strong enough. They don't have to go right in the corner; if there isn't enough access for tools and machine bolts, you can double the next stud and install the holdown to this post. Make sure you nail the studs together with 12d or 16d common nails every 4", avoiding where the holdown bolt holes will be drilled.

Position the holdown and mark all three holes.

A 90-degree angle drill allows for ease of horizontal drilling between bays or in tight areas. Hold the drill securely with both hands so that the bit is level. Select a wood bit that is 1/16" larger than the machine bolt diameter.

Here we are installing an expansion bolt, although you can also use threaded rod and epoxy.

Secure the machine bolts tightly with nuts and washers.

Here are two views of a properly installed holdown.

The holdown can also be installed above the plate; it doesn't have to sit flat on the block to work properly.

OTHER FOUNDATION TIES

Where there is not enough clearance above the mud sill to use the rotary hammer drill, you will not be able to install vertical bolts as we have illustrated. These alternative ties use horizontal drilling and bolting to connect the foundation and framing.

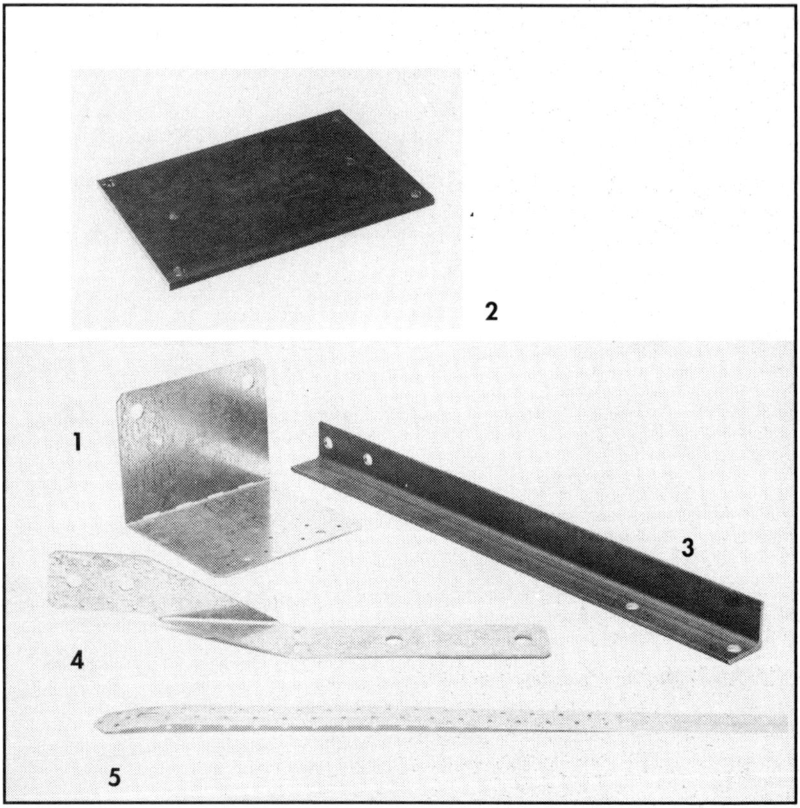

1. SIMPSON BRAND FA MUDSILL TO FOUNDATION TIE
2. 1/4" STEEL PLATE MUDSILL TO FOUNDATION TIE
3. ANGLE IRON JOIST TO FOUNDATION TIE
4. SIMPSON BRAND FSA JOIST TO FOUNDATION TIE
5. SIMPSON BRAND FLAT COIL STRAPPING

When there is insufficient room for vertical drilling as shown in our mock-up, you must drill horizontally into the face of the foundation. Typically, the length of the horizontal bolts should be about one-half the width of the foundation. For instance, a 12-inch wide foundation should have a 6" bolt. The spacing may vary from regular bolts; check your engineer's specifications. Use either epoxy capsules or cartridges when installing horizontal bolts.

This type of tie, like vertical anchor bolts, resists lateral movement. The horizontal plate is nailed to the mudsill, and the vertical plate is bolted to the foundation. When installing bolts within several inches of the top of the foundation, the stress created by expansion bolts may break the concrete. To avoid this, use threaded rod and epoxy.

In cases where the mudsill is flush with the inside face of the foundation, you may use 1/4" steel plates to tie the face of the mudsill to the foundation. You must use lag bolts to connect the plate to the mudsill.

Like holdowns, these ties prevent uplift. They tie the foundation directly to the floor framing. Use machine bolts to connect the tie to the framing.

Angle irons must be cut and drilled by a machine shop to your specifications. They do not come in standard sizes like the FA and FSA hardware.

These ties come in standard lengths with predrilled holes.

SHEAR WALLS

In the 1989 Loma Prieta earthquake many houses that were properly bolted still fell off their foundations. This is because the bolts do not tie the whole house down; they only connect the mudsill to the foundation. Another type of connection— shear walls — is needed to tie the foundation and mudsills to the cripple walls that support the house.

Shear walls are created by nailing plywood sheets onto the cripple walls. The sheets must extend from the mudsill to the top of the wall, and must be securely nailed. Follow your engineer's specifications for the thickness and grade of plywood.

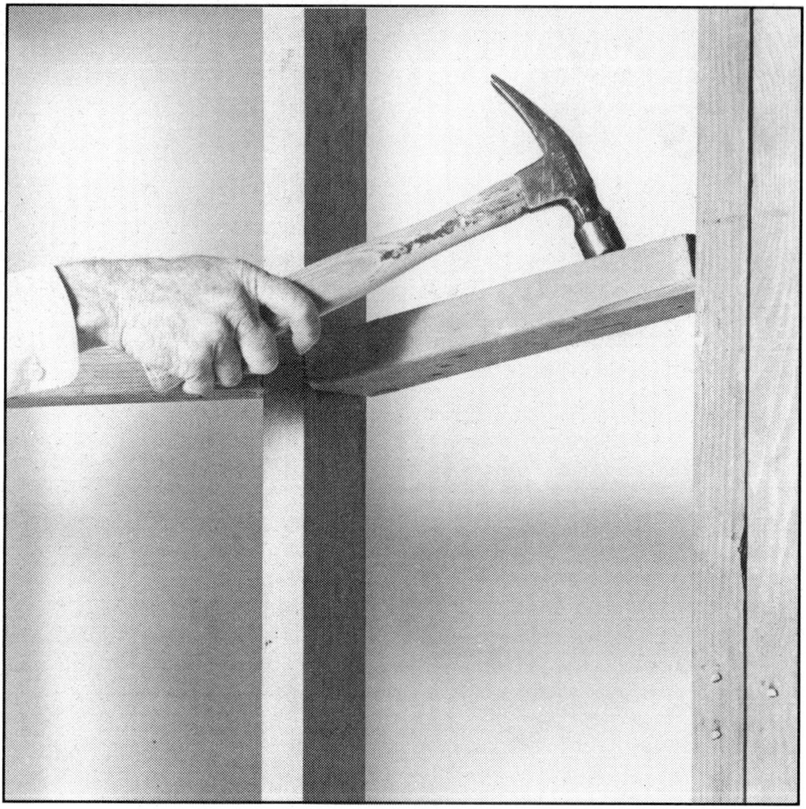

The most important principle to remember is that shear walls must be nailed on all four edges - the entire perimeter of each plywood sheet. There has to be solid framing behind each edge: studs for each vertical edge and either the mudsill and top plate or blocking behind each horizontal edge. Make sure the blocks fit tightly and are nailed securely with 16d common nails to the studs.

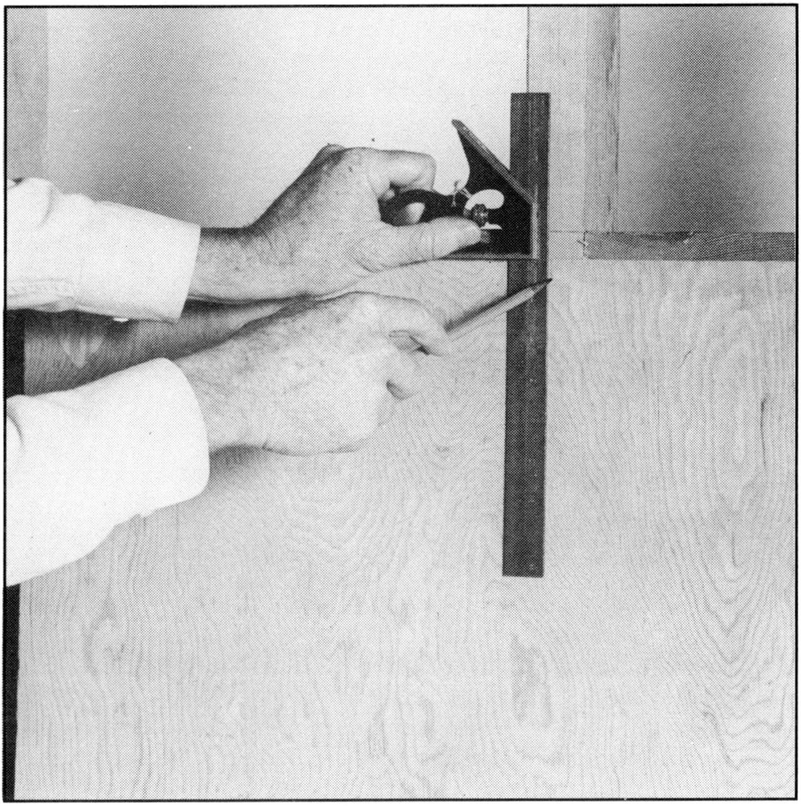

Plywood comes in 4' by 8' sheets. It can be installed either horizontally or vertically. Note how the blocking has been centered on the joint where two sheets meet so there is a nailing surface for each sheet.

While nailing the perimeter is crucial, it's also important to nail the plywood to all the studs behind the middle of the sheet. This is called field nailing.

Once the sheet is nailed up, the studs will be hidden and we won't know where to drive the nails. Here we have positioned the plywood and are marking the location of the studs on the sheet. Note that we are marking the center of the stud. The easiest way to mark the entire length of the sheet is to snap a chalk line. An alternative way is to measure the stud location from the corner of the wall and then mark the sheet accordingly.

Note that each edge of the sheet falls on the center of solid framing.

The quickest way to nail the shear walls is with a nail gun. You can rent the gun, air compressor and hoses. This is a powerful tool. Always wear eye protection and keep your hands away from the tip. The tip of the gun must be held firmly against the wood before the trigger can be pulled. When two sheets of plywood meet over a 2" x 4" stud or block, there is only 3/4" of nailing surface. Angle the gun slightly to ensure that the nails will hit the stud or block. Position the nails no closer than 3/8" from the edge. The nail head should be flush to the surface.

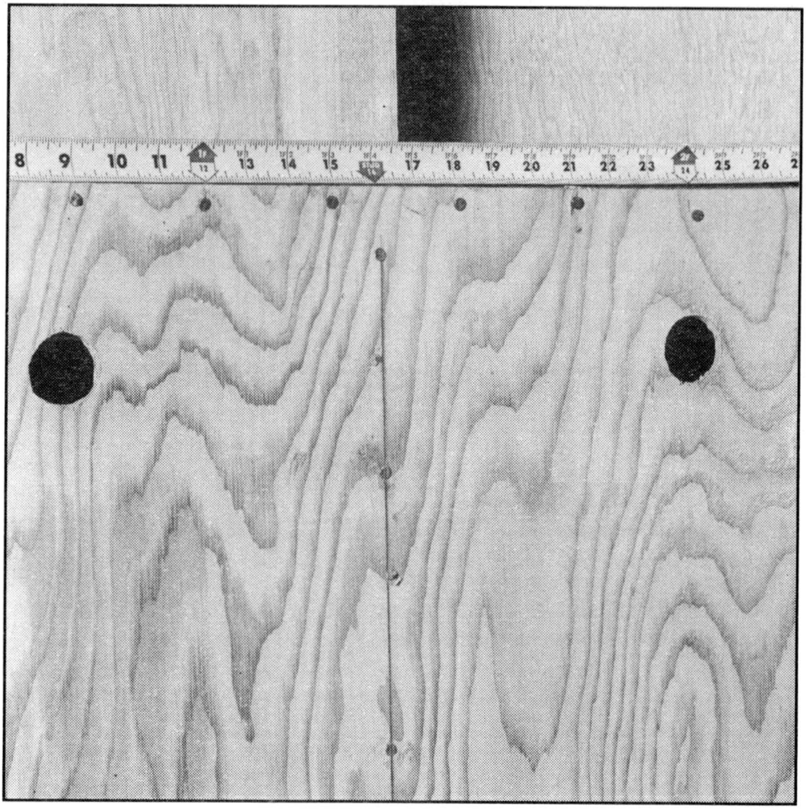

When nailing shear walls, use 10d nails spaced 4" apart on all edges, and 10d nails 6" apart in the field. The strength of the shear walls depends almost entirely on the quality and quantity of the edge nailing. Do not deviate from the nailing pattern and size specified by your engineer. Do not use drywall screws; they are not as strong as common nails. Note the chalk line denoting the stud location. Also note that studs are typically spaced 16" apart.

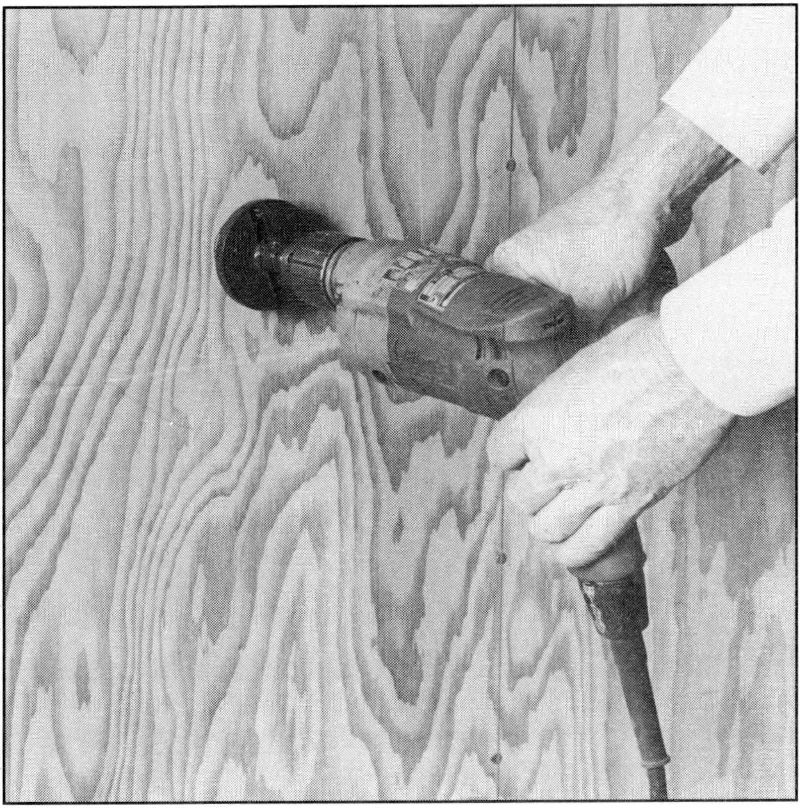

Moisture tends to collect inside walls close to the ground that are not ventilated. Drill 1-1/2" to 2" diameter holes in the plywood at the top and bottom of each stud bay to allow moisture to escape. Note that the holes are centered in each bay. It's easier to drill through plywood with a hole saw bit than with a flat bit.

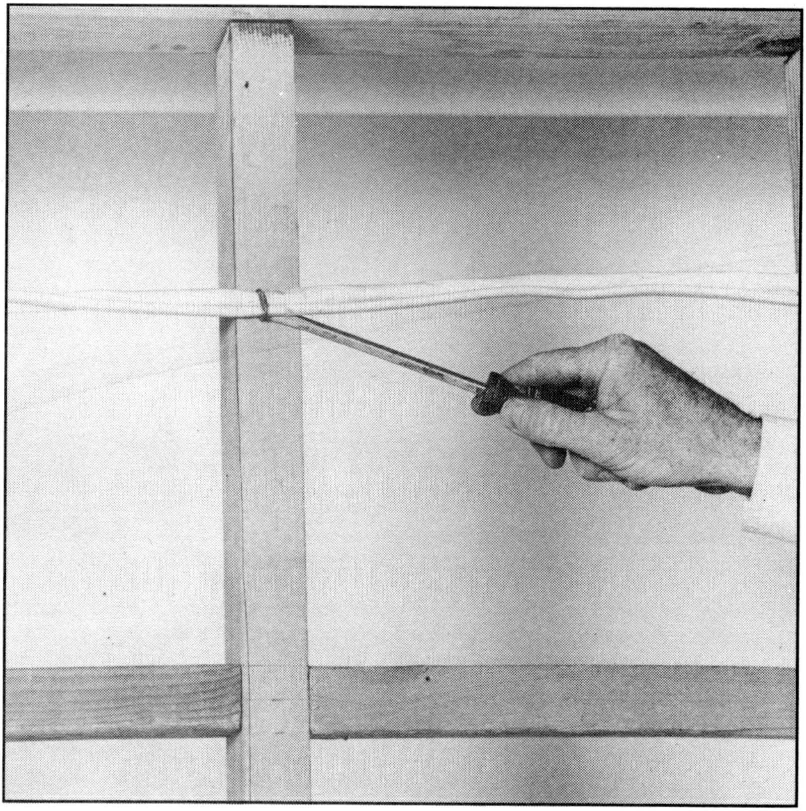

Sometimes you may find electrical wires or plumbing pipes are in the way. Remove the staples or brackets so you can slip the sheet of plywood behind the wires or pipes.

Be careful not to damage the wires or pipes when you slip the plywood underneath them. In some cases, you may have to call in an electrician or plumber to move wiring or plumbing. Be sure to reattach the wiring and pipes after you have nailed the plywood.

Large openings such as windows or doors weaken plywood shear walls, and they need to be reinforced with metal strapping that is nailed across the top and bottom of the opening. The strapping is nailed over the plywood, and extends beyond the edge of the opening. You need to install blocking on both sides to provide a nailing surface for these horizontal straps. This photograph illustrates an opening and the blocking that has been added on either side at the top; similar blocking must be added at the bottom of the opening.

The plywood has been nailed and the strapping has been installed. Descriptions of this type of reinforcement should be covered by your engineer's specifications.

You may need to notch around plumbing pipes or electrical boxes. Note blocking that has been installed for perimeter nailing. This notch needn't have been so large.

SHEAR WALL TO FLOOR TIES

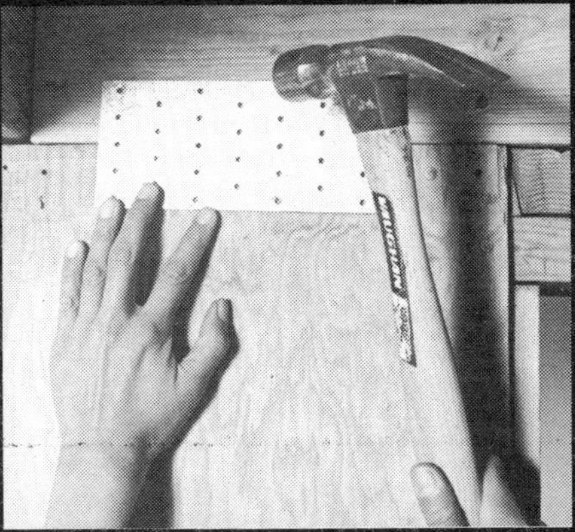

After bolting the mudsill to the foundation, installing holdowns to the corners, and tying the mudsill to the wall framing with plywood, the next important connection is between the shear wall and the floor framing. Even if the mudsill is bolted and the cripple wall has been strengthened with plywood, the floor can slide off the cripple walls in an earthquake if it has not been tied to the cripple walls. The floor framing can be tied down with a variety of metal ties. ❧

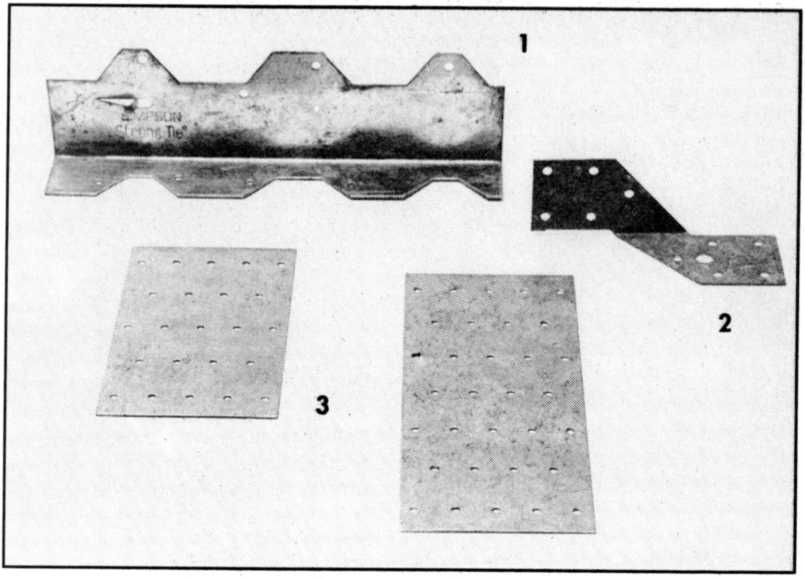

1. **90-DEGREE ANGLE TIE.**

This is installed on top of the cripple wall and ties the rim joist or blocking to the wall. (see p. 65)

2. **HURRICANE TWIST TIE.**

This connector is nailed into the floor joist and the top of the cripple wall. (see p. 72)

3. **FLAT NAILING PLATES.**

When the blocking between the floor joists is flush with cripple wall, you can use these plates to tie the wall to the blocking. (see p. 71)

All hardware shown is manufactured by Simpson Tie Company, Inc.

The L90 fits into the corner formed by the blocking and the top of the cripple wall. Your engineer's specifications should indicate the spacing of the ties.

The L90 tie sits on top of the cripple wall, so there isn't much room for nailing. Turning the hammer sideways allows for easier nailing in tight spaces.

Note the flat plate on the right. Some professionals recommend tilting it like this when you install it because putting tightly spaced rows of nails in a level plate tends to split the wood behind the plate.

A palm nailer drives single nails and can be used in tight clearances. Each nail is hand-placed in the tip; pushing down on the tool activates it. As always, hold the tool securely and wear eye protection.

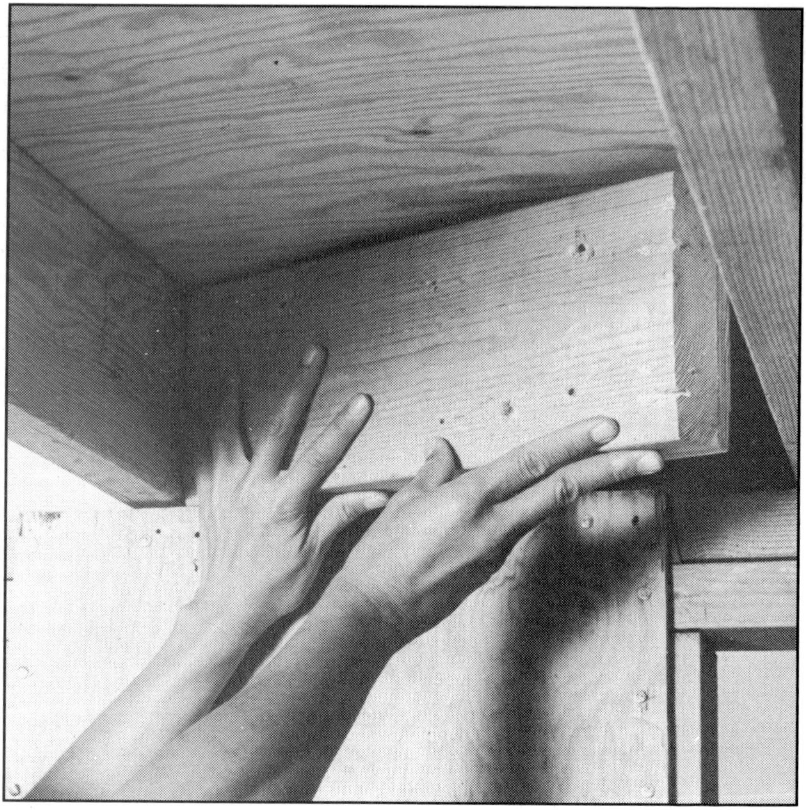

An alternative way to tie the floor framing to the wall is to add blocking between the joists so the face of the blocking and the face of the wall are flush.

Since standard 2 X 4 lumber is 3-1/2" wide but only 1-1/2" thick, you need to add a piece of wood to the block so that when it is installed it is flush to the edge of the wall.

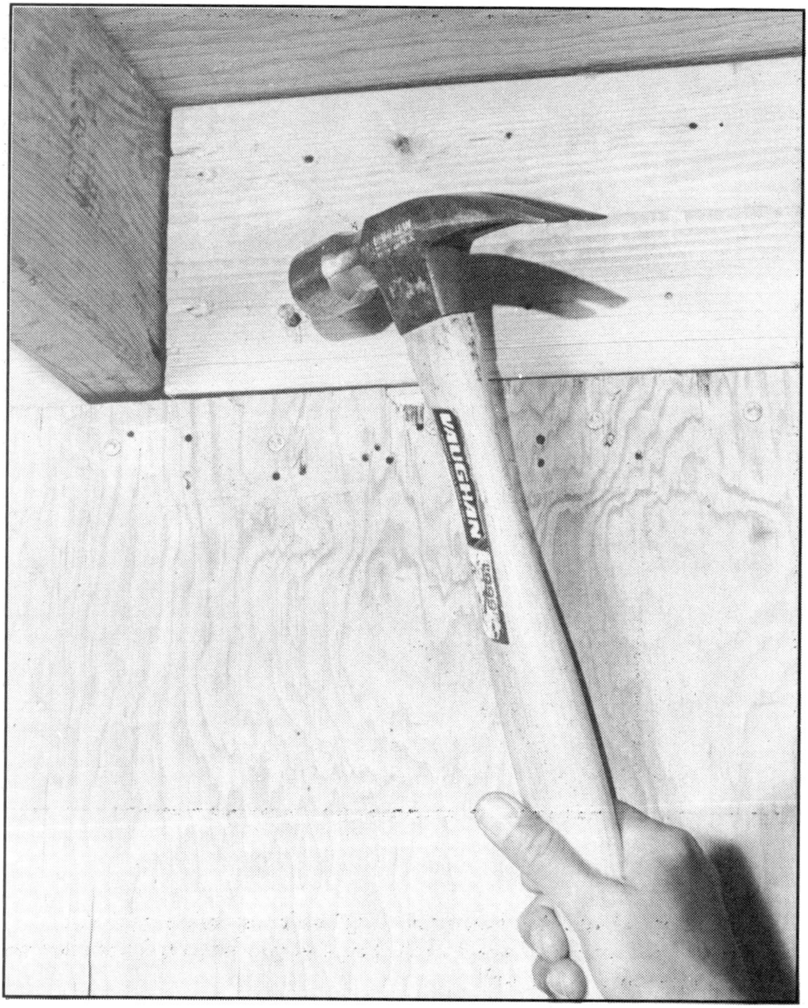

Nail the blocking into the joists as well as into the rim joist behind it.

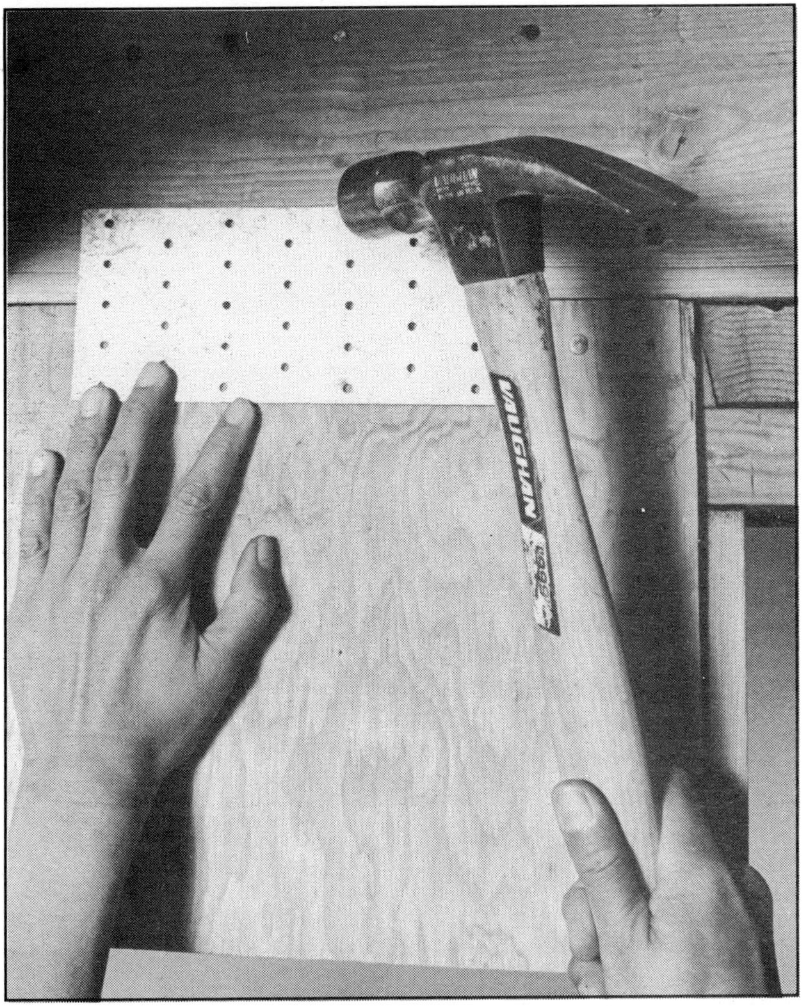

A flat plate ties the new block, and thus the whole floor framing, securely to the cripple wall.

Some contractors tilt the plate slightly so the nails won't be in a straight line with the wood grain. If the nails are in a straight line with the grain lines, they might split the wood.

Another commonly used connector is the Hurricane twist tie, which connects the joist directly to the cripple wall. Note that these connectors resist uplift, but are not designed like the L90 ties to resist lateral movement.

Larger versions of the twist tie can be installed on the exterior of the cripple walls in cases where the siding has been removed.

ADDITIONAL NOTES

1. Obtain all necessary building permits before starting construction.
2. Many professionals recommend using pressure-treated lumber (PTL) in all areas that are close to the ground or that sit on the foundation.
3. Not all materials are equal; there are many different grades of plywood and lumber and the structural strength of the material is often directly related to the grade. Follow your engineer's specifications.
4. Well-maintained, good quality tools make any job easier. The tools shown in the manual are professional quality and were selected as commonly used brands.
5. The quality of the retrofitting work depends on the connections. Follow the nailing pattern specifications: don't scrimp on nails in retrofitting work.
6. If the foundation of your house is not good quality reinforced concrete— for example, brick— have it inspected by an engineer. You may need to repair, strengthen or even replace the foundation.
7. Dryrot is typically caused by leaks or by poor drainage. Replacing the rotted framing is not a permanent solution; find and repair the cause of the leak or dampness.
8. There are also a number of ties that are designed to connect the floor and roof framing to the walls of the house. Consult with your engineer on which connectors are appropriate for your house.

ABOUT THE BUILDING EDUCATION CENTER

Building Education Center, 812 Page Street, Berkeley, CA 94710. 510/525-7610. Non-profit organization offers a Saturday class on "Earthquake Retrofitting Techniques for Homeowners." This one-day seminar is taught by Tony DeMascole and Jim Gillett, and uses a full-scale mock-up to demonstrate both retrofitting principles and techniques. Offered in Berkeley only. BEC offers 50-plus other classes on building and remodeling. To request a catalog, or order a copy of this book ($17.01 incl. tax & shipping), phone 510/525-7610 or FAX 510/525-0855.

OTHER RESOURCES

- **The Earthquake Project,** Northern California: Metro Center, 101 8th Street, Oakland, CA 94607, 510/893-0818; Southern California: 1110 East Green Street, Suite 300, Pasadena, CA 91006, 818/795-9055. The offices are funded by the State of California and have a wide variety of publications on seismic safety and structural strengthening. Call for a catalog of publications.

- **Applied Technology Council,** 3 Twin Dolphin Drive, Redwood City, CA 94065, 415/595-1542. Provides technical guides to structural strengthening for engineers, architects and contractors. Call for a catalog of publications.

- **California Seismic Safety Commission,** 1900 K Street, Suite 100, Sacramento, CA 95814, 916/323-4213. Distributes the 28-page *The Homeowner's Guide to Earthquake Safety* ($2.25 each).

- **Federal Emergency Management Agency (FEMA)** has a National Earthquake Hazards Reduction Program, which publishes the *Handbook for Seismic Rehabilitation of Existing Buildings* (FEMA-172) for engineers, architects and contractors. Order from FEMA, P.O. Box 70274, Washington, D.C. 20024.

- **Earthquake Engineering Research Institute (EERI),** 6531 Fairmount Avenue, Suite 7, El Cerrito, CA 94530, 510/525-3668. Provides technical guides to structural strengthening for engineers, architects and contractors. Call for a catalog of publications.

- **Local building departments** sometimes have information on structural safety projects and applicable local building codes.

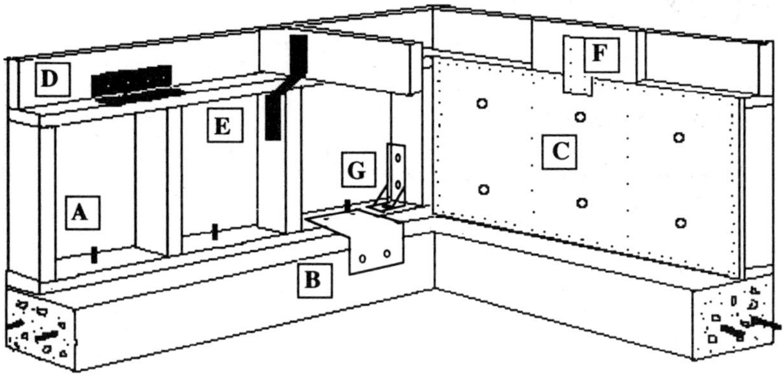

INTERIOR VIEW OF TYPICAL CRAWLSPACE

A. Anchor Bolts
B. Foundation to Mudsill Connector
C. Plywood* (making this a shear wall)
D. L-90 Wall to Floor Connector
E. Wall Stud to Floor Joist Connector
F. Flat Wall to Floor Blocking Connector
G. Holdown

* There are various grades and thicknesses of plywood. For retrofitting work, use 3/8" or 1/2" Structural I or Structural II grade plywood. The thickness you use should be determined by your engineer; thicker is not necessarily better.

Epoxy
metal cutting sawzall blades
latex gloves
rubber globe attach to rubber tube